BOOM
Fundamental Gravity as the potential source for the light refraction and diffraction and scattering and absorption

MOHSEN LUTEPHY

ISBN: 9798721596339

DEDICATION

On the fully variable light speed (VSL) universe which is derived by alliance of the Mach inertia principle and Planck's quantized natural units and generalized Minkowski metric (Einstein-Riemannian), it is extracted the equation of variable light speed. Then the light speed is varied point to point and fundamental gravity as a vector fulfills the light speed variations along the path via a tangential component and a component normal to the light velocity according to the Pythagorean Theorem and interestingly on the fundamental gravity we argue Snell's law of light refraction (BOOM!). Of course, the fundamental gravity does include bound quantum systems extended from Femto scale (Nucleuses and strong nuclear force), Micro-gravity (fundamental rainbow's gravity), the galaxies and clusters as the large scale bound quantum systems until to the observable universe which the variable gravitational G is quasi-Newtonian there (constant G in short cosmic interval of time). Refraction and diffraction of the light are sourced potentially by the fundamental Rainbow's gravity in bound quantum systems limited to the range about the wavelength of the photons. Quantum mechanically the photons are force carrier below the radius equal to their wavelength reasonable to enhance the gravitational G inasmuch as large that the atoms could to bend the photons highly (fundamental rainbow's gravity), similar to the enhanced gravity in the nucleuses by the hadron force-carriers which they cause to shape Femto scale bound quantum systems in the scale of the wavelength of the hadrons. The light's diffraction and scattering are also driven potentially by the gravity as the source of light bending in the light diffraction; of course the wave patterns are derived by the wave equation. In reality the edge of objects has the role of secondary wave sources for coherency and in the light scattering, the role of electrons is important for occurrences. Also we find that the path of photons in the diffraction despite the path derived by Huygens-Fresnel principle is not straight but the paths are curved, that is, the paths are quantized however interestingly the quantum mechanical path quantitation coincide the pattern driven by Huygens-Fresnel mathematical model. Also we find that absorption of the photons by particles is sourced by fundamental gravity and then the gravity has an essential role inn the thermodynamics.

CONTENTS

ACKNOWLEDGMENTS

This article is inasmuch as manifest and strong in arguments that we sure the scholars will distribute the topic in level of the book.

1 INTRODUCTION

 The light refraction is a mystery yet. Newtonian gravity and Einstein general relativity and Maxwell electromagnetism can't answer to the question that why the light is refracted? It is not clear still why the light speed is slowed in the mediums?

We know that the explanations based upon the idea of light scattering from, or being absorbed and re-emitted by atoms are both incorrect. Explanations like these would cause a "blurring" effect in the resulting light, as it would no longer be travelling in just one direction.

As mentioned in the Wikipedia dictionary, ultimate explanation for light refraction has been proposed on the nature of the light as an electromagnetic wave. On this explanation, the light is an oscillating electrical/magnetic wave and the light traveling in a medium causes the electrically charged electrons of the material to also oscillate. Then it is thought that the oscillating electrons emit the electromagnetic waves of its own by this oscillation. It is thought that the emitted electromagnetic waves interact with the incident photons and when the waves interfere in this way, the resulting "combined' wave may have wave packets that pass an observer at a slower rate. The explanation states (Wikipedia) that:

"if it reaches the interface between the materials at an angle one side of the wave will reach the second material first, and therefore slow down earlier. With one side of the wave going slower the whole wave will pivot towards that side. This is why a wave will bend away from the surface or toward the normal when going into a slower material. In the opposite case of as wave reaching a material where the speed is higher, one side of the wave will speed up and the wave will pivot away from that side."

This explanation is unjustified; for imaginary interfere of the incident photons with produced electromagnetic waves by atoms and actually there is no such a chance and there is no any report to reproduce it in the laboratory. Also if the wave packet was the source to decrease the speed of light, then the white light which is also a wave packet had slower speed compared to the monochromatic beams. Also the change of angle is too a big difficulty for this explanation for light refraction. The slowing of a part of the photon which reaches earlier to the different medium has no any mechanical source for change of the photon direction in the interface of the mediums and disintegration of these fictitious wave packets to pure waves, while the waves come out from the medium is also a paradox. Generally this explanation has no any argument for the Snell's law and slowing of the photon speed in the mediums by the group velocity of the wave packets has no any link to the Snell's law. The refractive index is an experimental index and up to now there is no a mechanism to illustrate it and we don't know why the refractive index is varied in different mediums and all are assumptions whereas the Snell's law is very verse and such a law requires a verse origin.

The light is neutral electrically and then electromagnetic force can't refract the light. Also the Fermat least time principle and Huygens-Fresnel principle don't indicate the mechanical source for the light speed variation suppose these principles are technical lemmas.

The argument for the light refraction on the theory of envelope mechanism and group velocity is mistake for many reasons for example the light is refracted when the light is even still not completely interred into the mediums. There should be a mechanical force affecting the photons from side of the atoms. The electromagnetic force can't be the case for the reality that the photons are mainly electrically neutral.

Newtonian gravity is not fundamental because that the newton's gravity is an additional effect of the multi-particle system assumed as a single fundamental mass whereas that additional effect of individual particles may show a different property which is not fundamental but composite mistakenly assumed as a fundamental property. Newton's gravity is approximate. Even here we will show that the fundamental gravity is not centrifugal force but additionally all together in a many-particle static body behaves as a centrifugal resultant force. We show here that the fundamental gravity does work individually for a mass-point particle and then the force on an apple is not fundamental and what we see as a resultant gravity may differ with the fundamental gravity acting on the fundamental particles.

Also the Einstein general relativity isn't a fundamental gravity because that Einstein total field is a proof less formula and consequently modified and many mathematical answers can be obtained by that. The scientists for each puzzle; they try to find a solution from Einstein general relativity to set it with the observations and when they can't find a suitable answer they consider modifications to the general relativity or its solutions. Then they have assumed unobserved mass/energy and undiscovered sub atomic particles for interpretation of observations.

By the way there are many evidences revealing that the Einstein general relativity is not fundamental, for example one of the evidences is that in the Einstein general relativity, it is used a parameter defined as the volume density of the universe whereas that the universe isn't homogenous and principle of cosmology isn't valid. In general relativity, the energy tensor $T_{\mu\nu}$ is proportional to Riemannian curvature $R_{\mu\nu}$ but in the vacuum or voids, the energy tensor $T_{\mu\nu}$ is zero whereas the curvature $R_{\mu\nu}$ is not zero as a contradiction for GTR. In reality discontinuity of the density is manifest and there is no any volume included to the uniform distribution of the matter inside, whether large scale or small scale. Then every equation included to the volume density is not fundamental for that the volume density is approximate and approximation is not fundamental.

On the Einstein general relativity, the scientists have used the vacuum solution to simulate the solar system whereas in the solar system, the universal potential is large and it is false to simulate the solar system independently which mistakenly assumed in the Schwarzschild solution for the light passing around the Sun. Also the GTR has answers for vacuum as a manifest contradiction because that the Einstein by Mach inertia principle has claimed that general relativity should be fully on the curvature of space by matter only (Pais, 2005).

On the Mach inertia principle, the mass is not an intrinsic property of the matter until to be defined locally but the mass is an extrinsic property driven by whole of the universe matter in the Machian relation that

$$m = m_0 \frac{G_N}{c_0^2} \sum_i \frac{m_i}{r_i} \tag{1}$$

Then the approximate equations defined on volume density should be redefined on the parameter $\sum_i m/r$, that is, we need to revise the energy tensor on the parameter $\sum_i m/r$.

Einstein total field solutions encounters next paradoxes such as the anisotropy of the light speed by direction, or the existence of the singularities or a region inside the Schwarzschild radius which the scientists have considered it as a black hole. In fact the Einstein general relativity is a continuum frame work mechanics and not possible to apply it precisely in N-body problem properly. Such a continuum mathematical frameworks are indeed two body problems. Even in the Einstein especial relativity, we see that the N-body problem of the Maxwell electromagnetism encounters with relevant paradoxes revealing that ultimately the continuum framework can be an approximation for real N-body discrete universe. Then Einstein general relativity is an approximate model, by applying a continuum framework for discrete physics. There are many logical failures in the Einstein general relativity all revealing that the Einstein general relativity is an approximate mathematical model for the gravity.

But the fundamental gravity is agreement with generalized Minkowski metric which the light speed c is fully variable on the Mach inertia principle. Then the curvature of the space depends exclusively to the light speed and light speed in the universe does determine the curvature fully. Here we find fundamental gravity which reduces to the Newtonian gravity approximately and even mimics GTR. We see that the fundamental gravity and refraction of the light are unified to indicate that the refractive index depends fundamentally to the gravitational potential instead the volume density and we find a true correlation between the refractive index and the parameter $\sum_i m_i / r_i$ in bound quantum systems which we call it here absolute density.

Up to now the idea of scientific consensus is that the light refraction is a local event and in the vacuum the refractive index is at the order of unity however the light speed is variable vacuum to vacuum in different points of the universe. It is believed that the refractive index is an electromagnetic parameter driven via Maxwell electromagnetism. But reestablished Newtonian mechanics yields to the fundamental gravity which is identical with refraction of the light and the mechanical potential of the light refraction is the gravity. We argue here the so called Snell's law of the light refraction by fundamental gravity and we find a precise answer for question why an object is crystal and why other not.

The fundamental gravity is revealing the true rainbow's gravity which is possible to be written in metric format too. The previous rainbow's gravities (Galan and Marugan, 2004., Hackett, 2006., Garattini and Mandanici, 2012., Garattini and Majumder, 2014., Garattini, 2013., Garattini and Majumder, 2014., Leiva et al. 2009., Li et al. 2009., Ali et al. 2015., Awad et al. 2013., Barrow and Magueijo, 2013., Liu and Zhu, 2008., Ali and Khalil, 2014., Gim and Kim, 2014., Amelino-Camelia et al. 1997., Amelino-Camelia et al. 1998) on the principle of relative locality (Amelino-Camelia, Giovanni, et al., 2011) suggest that the gravity affects different wavelengths in the same way that a prism affects the light. The theory was first proposed in 2003 by physicists Lee Smolin and Joao Magueijo and scientists are currently attempting to detect rainbow's gravity using the Large Hadron Collider (Banks and Fischler, 1999., Giddings and Thomas, 2002., Dimopoulos and Landsberg, 2001., Emparan et al.. 2000., Meade and Randall, 2008., Antoniadis et al. 1998., Rocha et al. 2006).

But here we find that the light refraction and the gravity are unified and the electromagnetism isn't the source for the light refraction and derivation of the refractive index by Maxwell electromagnetism is imaginary suppose the photons are electrically neutral and we need to reestablish the physics. According to different phenomenological motivations, a series of RG (Rainbow's Gravity) models have been obtained. Based on the varying speed of light theory, Magueijo and Smolin proposed a kind of MDR (Modified Dispersion Relation) as

$$\frac{E^2}{\left(1-\gamma E l_p\right)^2} - p^2 = m^2 \qquad (2)$$

Where γ, l_p, and m represent the rainbow parameter, the Planck length , and the m, mass of the test particle. This equation indicates that the space-time has an energy-dependent velocity as

$$c(E) = dE / dp = 1 - E l_p \qquad (3)$$

Comparing with general form of MDR it has been deduced that

$$E^2 f^2\left(E l_p\right) - p^2 g^2\left(E l_p\right) = m^2 \qquad (4)$$

And from Feng and Yang (2018) we have

$$ds^2 = -\frac{1-(2GM/r)}{f^2\left(E l_p\right)} dt^2 + \frac{dr^2}{\left[1-(2GM/r)\right]g^2\left(E l_p\right)} + \frac{r^2}{g^2\left(E l_p\right)} d\Omega^2 \qquad (5)$$

But we indicate here that the light speed doesn't depend to the energy of photon but all photons have equal speeds in the vacuum in each point of the universe and we argue that assumed rainbow's parameter γ is zero and still correct the energy-momentum relation $E^2 - P^2 = m^2$.

The general relativistic rainbow's gravities have no answer for refraction of the light but noticeable for bending of the light around the Schwarzschild radius in modified general relativity whereas we argue here that the fundamental gravity (reestablishment of the Newton's gravity) is the potential source for the light refraction and the atoms bend the light extremely much inasmuch as that the atoms simulate what scientists are thinking to happen around the fictitious black holes. Of course the fundamental gravity does not show any relativistic black hole and ultimately the answer is just spiral geodesy in agreement with experiments performed at LHC with no result for black holes (Chatrchyan, et al. 2012a, 2012b).

We find here also a metric tensor which mimics dependency of the fundamental gravity to the photons wavelength and interestingly we find that some metric tensors in the field of general relativity in extra dimensions mimic the bound quantum systems and general relativity may be generalized to the bound quantum systems like the Newtonian gravity (Lutephy, 2020).

Of course, we don't agree with deformation of the energy-momentum relation applied in relativistic rainbow's gravities. Also we don't agree with extra dimensions applied by rainbow's gravities in the general relativity; however such extra dimensions cause to extend mathematically the solutions as a degree of freedom to embed approximately the bound quantum systems in the general relativity. But a mathematical model does not show ever a true physical mean and we agree the criticism of Sabine Hossenfelder for rainbow's gravities in title "No, the LHC will not make contact with parallel universes" (backreaction.blogspot.com 2015).

Our results here can be modeled approximately via a generalized Schwarzschild metric which the metric tensor depends to the wavelength of the photons in the way of quantization of boundary in the scale comparable to the wavelength of photons which causes the geodesy of photons to be different by different wavelengths, of course without relation to extra dimensions and parallel universes, verifying the Sabine Hossenfelder statements in her criticism.

We show here that the change of the light speed does relate to the wavelength of the photons for reality that the wavelength of photons does produce quantum mechanically a bound quantum system below the radius equal to the wavelength of the photon and then it is revealed high intensity gravity for limitation of the potential to the boundary of bound quantum systems, exactly similar to the strong gravity observable in the strong nuclear force in which the wavelength of the hadron force-carriers produce the bound quantum systems in the nucleuses (Lutephy, 2020), or similar to the gravity in the galaxies and clusters which they don't obey the Newtonian and Einstein gravities. As argued by Lutephy (2020), the galaxies and clusters are large scale bound quantum systems which the potential is limited to the boundary of galaxies and clusters and then we see very different observations mistakenly concerned to an unobserved thing called dark matter. It is funny that the scientists have calculated the dark matter profiles to illustrate the failure of the Newtonian and Einstein gravities whereas that this is inverse engineering of an imaginary thing.

We have argued here that the change of potential boundary causes to vary the size of gravity but not its formula fundamentally. We have gravity in different systems, not different gravities.

By the way we find that the Newtonian gravity is generalized to the quantum bound systems related to the wavelength of the force carriers or wavelength of the test mass itself as a self-force carrier. Then the gravity in the galaxies and clusters as the large scale quantum bound systems and gravity in the nucleuses as the strong nuclear force and true rainbow's gravity all are identical and unified.

Relativistic Rainbow's gravity scientists believe that the metric components depend to the energy of the test mass but they have not proposed a unified mechanism for refraction and gravity. But here we indicate that the refraction is the gravity and even we indicate that the mechanical potential of the light diffraction is the fundamental gravity. Of course we will see that the size of the gravity is not alone factor for differences in different systems but the fundamental gravity is varied in the vector too and even we see that the fundamental gravity is not directed to the side of the center against the Newtonian gravity suppose it mimics the Einstein general relativity rather than the newton's gravity. The fundamental gravity is self-existence so that via assuming some approximations transferred to the Newton's gravity and via other assumptions approximately transferred to the Einstein Gravity in the level of the rainbow's gravity too. Then the fundamental gravity is not indeed a modification for Newtonian and Einstein's gravities but here we return to the fundamental gravity independently and reproduce approximate models hereafter.

2 SNELL'S LAW VIA THE FUNDAMENTAL GRAVITY

Here the base of the gravity is the light velocity in fully variable light speed (VSL) universe (Lutephy, 2019) and we have

$$c = \frac{\Diamond}{\sum_i \frac{m_i}{r_i}} \qquad (6)$$

Where c is the photon speed and m is the mass of the particles and r is distance of the particles from the photon and $\Diamond$ is a universal constant and we call this equation here the Alpha Genesis Prime.

This equation is derived by Alliance of the Mach inertia principle and Planck's quantum natural units and generalized Minkowski metric.

Newtonian mechanics is inconsistent with VSL for that it is ever attraction and does not obey exclusively dependency of the light speed to the gravitational potential.

But here we find the fundamental face of the Newtonian mechanics for its generality in N-body problem in fully VSL universe.

The newton's gravity should be reorganized on the Alpha-Genesis-Prime (Eq. 6). This means that the newton's force has a component ever tangent on the light velocity. In N-body problem, the role of tangential component is to justify the variation of light speed via Newtonian type force.

When the light is moving in the space of N-body universe, then the light speed is varied by Alpha Genesis Prime (Eq. 6) and then equivalently we can consider a vector force from the N-bodies in which the variation of light speed by them is additionally equal to what the equation of Alpha Genesis Prime does show it.

Then we have ever a tangential component that

$$f_{i\vec{c}} = \frac{Gm_i}{r_i^2} \frac{\vec{r} \cdot \vec{c}}{|\vec{r}||\vec{c}|} \qquad (7)$$

The Alpha Genesis Prime should be confirmed by this force along the path of the light in the N-body space (universe).

By Alpha Genesis Prime (Eq. 6) it is deduced that

$$\Delta c = -\frac{\Diamond}{\Theta^2} \Delta\Theta \qquad (8)$$

Where we define a parameter called here absolute density Θ that

$$\Theta = \sum_i \frac{m_i}{r_i} \tag{9}$$

We have

$$\Delta\Theta = -\sum_i m_i \frac{\Delta r_i}{r_i^2} \tag{10}$$

Then from equations (8, 10) we deduce

$$\Delta c = -\frac{\Diamond}{\Theta^2} \sum_i \frac{m_i \Delta r_i}{r_i^2} \tag{11}$$

We know mathematically that

$$\Delta r_i = \Delta r \frac{\vec{r} \cdot \vec{c}}{|\vec{r}||\vec{c}|} \tag{12}$$

Then from equations (11, 12) we obtain

$$\Delta c = -\frac{\Diamond}{\Theta^2} \sum_i \frac{m_i}{r_i^2} \frac{\vec{r} \cdot \vec{c}}{|\vec{r}||\vec{c}|} \Delta r \tag{13}$$

Also by (8) via the Newton's second law we have

$$\frac{G m_i}{r_i^2} \frac{\vec{r} \cdot \vec{c}}{|\vec{r}||\vec{c}|} = \frac{\Delta_i c}{\Delta t} \tag{14}$$

Where $\Delta_i c$ is the change of the light speed by i-th particle with mass m_i.
From (6) and (11) we deduce that

$$\Delta c = -\frac{\Diamond}{\Theta^2 G} \sum_i \Delta_i c \frac{\Delta r}{\Delta t} \tag{15}$$

Substituting $c = \Delta r / \Delta t$ into the above equation yields to

$$\Delta c = -\frac{\Diamond c}{\Theta^2 G} \sum_i \Delta_i c \tag{16}$$

Clearly we have

$$\Delta c = -\sum_i \Delta_i c \tag{17}$$

And then from equations (16, 17) it is deduced that

$$G = \Diamond \frac{c}{\Theta^2} \tag{18}$$

Substituting equations (6, 9) in the equation (18) we obtain

$$G = \frac{c^3}{\Diamond} \tag{19}$$

Also we obtain that

$$\Diamond = \frac{c^3}{G} \tag{20}$$

And constancy of c^3/G has been verified previously via alliance of the generalized Minkowski metric and Planck quantum natural units and Mach's mechanics (Lutephy, 2019).

Then from equations (6, 20) for a mass-point affected gravitationally by a mass like the sun with size M in a distance R in two body problem universe we have that

$$c = \frac{c_0}{\dfrac{G_N}{c_0^2}\sum_i \dfrac{m}{r} + \dfrac{G_N}{c_0^2}\dfrac{M}{R}} \tag{21}$$

Then from Machian relation $G_N \sum m/r = c^2$ we find that

$$c = \frac{c_0}{1 + \dfrac{G_N M}{c_0^2 R}} \tag{22}$$

$$c \cong c_0 \left(1 - \frac{2G_N M}{c_0^2 R}\right) \tag{23}$$

And this is identical with the light speed derived in the Schwarzschild metric. Eq. (23) is an evidence to indicate that the fundamental gravity is not centrifugal force but when the light is moving radial toward the mass M; the speed of the light is decreased despite the Newton's gravity. Of course failure of the Schwarzschild metric is that at the infinity, the light speed is c_0 whereas that here the fundamental gravity is fully VSL, that is, the light speed is determined by the matter only in the definition of the Mach inertia principle.

By the way from equations (7, 20) we find

$$f_{i\bar{c}} = \frac{m_j c^3}{\Diamond r_i^2} \frac{\vec{c} \cdot \vec{r}_i}{|c||r_i|} \tag{24}$$

This is tangential component of the fundamental gravity which its duty is to justify the light speed on the line of the Alpha Genesis Prime.

In the vector calculus, each vector is divided to the tangent and normal (perpendicular) components on the Pythagorean Theorem and then the complete vector of the fundamental gravity is drawn as

$$f_i = \frac{m_i c^3}{\Diamond r_i^2}\left[\frac{\vec{r}_i \cdot \vec{c}}{|\vec{r}_i||\vec{c}|} - \frac{\vec{r}_i \cdot \hat{c}}{|\vec{r}_i||\hat{c}|}\right] \qquad (25)$$

Where $\hat{c}$ is assumed as a vector in the size of the light speed, normal to the light velocity vector $\vec{c}$ so that $\hat{c} \times \vec{c} = c^2$.
Equivalently we can write that

$$f_i = \frac{m_i c^3}{\Diamond r_i^2}\left[\cos\varphi\frac{\vec{c}}{|\vec{c}|} - \sin\varphi\frac{\hat{c}}{|\hat{c}|}\right] \qquad (26)$$

Where φ is the angle between vectors, light velocity $\vec{c}$ and the distance $\vec{r}$.
Or we can write that

$$f_i = \frac{Gm_i}{r_i^2}\left[\cos\varphi\vec{M} - \sin\varphi\vec{N}\right] \qquad (27)$$

Where $\vec{M}$ is a unit vector tangent to the light velocity and $\vec{N}$ is a unit vector normal to the light velocity. Or we can say that the generalized Minkowski metric while the light speed is fully variable by Alpha Genesis Prime is the fundamental gravity in the language of the metric tensor.
As we see the fundamental force here is not centrifugal for that the sign of the normal component $\vec{N}$ is negative and this negative sign has a very important role. If the sign of the normal component was positive, the force was identical with Newtonian gravity and the light speed was increasing radially by force towards the active mass whereas that according to the Alpha Genesis Prime, the light speed is decreased similar to the Schwarzschild metric. In reality the fundamental gravity seems a Newtonian mode of the Einstein general relativity.
Then negative sign of the normal component is logical result of the Alpha Genesis Prime or generally the result of every kind of VSL which the light speed is increased by distance.

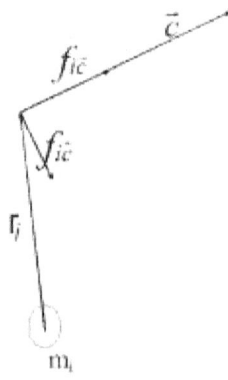

Figure 2. the simple shape of the fundamental gravity vector

In vector mode, the reestablished Newton's law, is the same Newton law but just the sign of the normal component of the vector $\vec{r}$ on the light velocity is reversed.

According to the Newton's second law and equation (27) we obtain

$$\frac{mc^3}{r^2}\cos\varphi = \frac{dc}{dt} \tag{28}$$

$$\frac{mc^3}{r^2}\sin\varphi = \frac{cd\theta}{dt} \tag{29}$$

In the x-y coordinate system mathematically we have

$$\varphi = \theta - arctg(dy/dx) \tag{30}$$

Substituting equation (30) into the equations (28, 29) it is deduced that

$$\frac{mc^3}{r^2}\cos(\theta - arctg(dy/dx)) = dc \tag{31}$$

$$\frac{mc^3}{r^2}\sin(\theta - arctg(dy/dx)) = cd\theta \tag{32}$$

Then we deduce that

$$\frac{dc}{c} = \cot(\theta - arctg(dy/dx))d\theta \tag{33}$$

In the additional form for a material surface we have

$$dy/dx = 0 \tag{34}$$

And then from equations (33, 34) we deduce that

$$\frac{dc}{c} = \cot(\varphi)d\varphi \tag{35}$$

Integration on the equation (35) results

$$\ln c = \ln|\sin\varphi| \tag{36}$$

Then it is deduced that

$$c\sin\varphi = c'\sin\varphi' \tag{37}$$

And this is the Snell's law of the light refraction, argued here by fundamental gravity.

For straightforward argument consider a material surface which the light passes it at an angle φ from y axis. The vector $\vec{r}$ is ever along the y-axis and then we have that

$$d\varphi = d\theta \tag{38}$$

Substituting equation (38) in the equations (28, 29) it is deduced

$$|f|\cos\varphi = \frac{dc}{dt} \qquad (39)$$

$$|f|\sin\varphi = \frac{cd\varphi}{dt} \qquad (40)$$

And the solution of this apparatus is clearly the Snell's law (Eq. 37).
Also from fundamental gravity (Eq. 26) we obtain

$$|f_i|r_i^2 = \frac{m_i c^3}{\Diamond} \qquad (41)$$

In reality we have

$$\frac{m_i c^3}{\Diamond r_i^2} = \frac{dc}{dt} \qquad (42)$$

$$\frac{m_i c^3}{\Diamond r_i^2} = \frac{c^2 d\varphi}{dl} \qquad (43)$$

Where dl is the differential arc of the path.
By the way substituting $d\varphi / dl = 1/R$ into the equation (43) we obtain

$$\frac{m_i c}{\Diamond r_i^2} = \frac{1}{R} \qquad (44)$$

$$\frac{m_i}{r_i \sum_j \frac{m}{r}} = \frac{r_i}{R_i} \qquad (45)$$

This is a version of the fundamental gravity and interestingly we see that the
equation is completely non scale.
And as argued in the book "non scale mechanics" (Lutephy, 2019), the
equation (45) is deduced from the below fundamental equations that:

$$|f_1|r_1^2 = |f_2|r_2^2 = \cdots = |f_n|r_n^2 = c^3 \qquad (46)$$

$$|f_1|r_1 + |f_2|r_2 + \cdots + |f_n|r_n = c^2 \qquad (47)$$

The Equation (46) does show that the inverse square law of the Newton is
relative and against the Newton dynamics, there is no a universal constant
G that $fr^2 = Gm$ but the correct sentence is relative sentence in the N-
body universe that $fr^2 = f'r'^2$.
In realty from equations (6, 43) we have

$$\frac{c^2 d\varphi}{dl} = \frac{m_i c^2}{r_i^2 \sum_i \frac{m}{r}} \qquad (48)$$

And this relation reduces to the equation (47) when the potential integration is overall on the whole observable universe.
Then it is deduced that

$$\frac{m_i c^2}{r_i^2 \Theta} = a \tag{49}$$

Or equivalently

$$\frac{G_N m_i}{r_i^2} = \frac{G_N \sum\limits_i \frac{m}{r}}{c^2} a \tag{50}$$

And this equation is what we see it in the galaxies and clusters versus the fictitious dark matter (Lutephy, 2020).
We can follow on the metric method too. Via generalized Minkowski metric (Lutephy, 2019) we find that

$$g_{00} = -c^2 \tag{51}$$

In relativistic mode (Lutephy, 2019) we have

$$c^2 = \frac{1}{\chi_2 \sum \frac{m}{r}} \tag{52}$$

Where

$$\chi_2 = G / c^4 \tag{53}$$

Then we have

$$g_{00} \sum \frac{m}{r} = -\chi_2^{-1} \tag{54}$$

From the geodesy equation we have

$$\frac{\partial^2 x_\mu}{\partial t^2} = \frac{\partial}{\partial x_\mu} g_{00} \tag{55}$$

Substituting $\Theta = \sum m / r$ in the equation (54) and then substituting g_{00} from equation (54) in the equation (55) it is deduced that

$$\frac{\partial^2 x_\mu}{\partial t^2} = \chi_2^{-1} \frac{\partial (1 / \Theta)}{\partial x_\mu} \tag{56.1}$$

$$\frac{\partial^2 x_\mu}{\partial t^2} = \frac{c^4}{G^2 \Theta^2} \frac{\partial (G\Theta)}{\partial x_\mu} \tag{56.2}$$

$$\frac{\partial^2 x_\mu}{\partial t^2} = \frac{\partial \varphi}{\partial x_\mu} \tag{56.3}$$

And the equation (56.3) is Newtonian gravity derived by fundamental gravity in metric mode.

Of course from equation $\varphi = c^2$ and substituting into the equation (56.3) we obtain that

$$\frac{\partial^2 x_\mu}{\partial t^2} = \frac{\partial c^2}{\partial x_\mu} \qquad (57)$$

Also for geodesy it has been defined that

$$\delta \int ds = 0 \qquad (58)$$

But for light the ds is ever zero and then this equation doesn't work for light suppose by this equation we obtain the geodesy of a body moving asymptotically closed to the light.

For precise light speed we have

$$dt = dl / c \qquad (59)$$

This equation is on the equivalency of the time and light speed.

The dt has the role of ds while the time is not curved. The time dimension indicates the light speed in each point of the place and independent definition of the time and light speed is one of the mistakes of the Einstein. If the time is curved then naturally the light speed should be ever constant whereas that in Einstein general relativity, the light speed is variable independently.

By the way when the curvature of the space is fully driven by VSL, we can find that the least path is the path that

$$\delta \int \frac{dl}{c} = 0 \qquad (60)$$

And then from equations (58, 60) it is deduced that

$$\delta \int dt = 0 \qquad (61)$$

And this equation verifies the Fermat's Principe of minimum time. And then the Fermat least time principle does violate the Einstein theorem in his general relativity for the equation of the geodesy that $\delta \int ds = 0$.

Interestingly the Fermat's principle of the least time, it results the Snell's law of the light refraction, verifying our results here.

3 FUNDAMENTAL GRVAITY AND EQUATION OF THE LIGHT PATH IN POINT-MASS UNIVERSE

And also we can calculate geodesy of the light in a point-mass universe comparable to the Schwarzschild solution.
From equations (28, 29) we obtain

$$-\frac{dc}{c} = ctg\,\varphi d\theta \tag{62}$$

And for point-mass universe where the test mass is mass point we have

$$c = \lozenge\,\frac{r}{M} \tag{63}$$

Then we have

$$-\frac{dr}{r} = \cot g\varphi d\theta \tag{64}$$

In x-y coordinate system, because the equations (64, 65) are transferable to each other, the equation (Eq. 64) is equivalent with evident equation below

$$\theta - arctg(dy/dx) = \varphi \tag{65}$$

Problem is that in the x-y coordinate system, the angle between the vector $\vec{r}$ and velocity of the light $\vec{c}$ is not exactly equal to φ but we have

$$\langle\vec{r},\vec{c}\rangle = \pi - \varphi \tag{66}$$

Then in the x-y coordinate system we have

$$-\frac{dr}{r} = tg\,\varphi d\theta \tag{67.1}$$

$$-dr/r = tg\left(\theta - arctg\,\frac{d(r\sin\theta)}{d(r\cos\theta)}\right)d\theta \tag{67.2}$$

$$-dr/r = tg\left(\theta - arctg\,\frac{\sin\theta dr + r\cos\theta d\theta}{\cos\theta dr - r\sin\theta d\theta}\right)d\theta \tag{67.3}$$

$$-dr/r = \left[\frac{tg\theta - \frac{\sin\theta dr + r\cos\theta d\theta}{\cos\theta dr - r\sin\theta d\theta}}{1 + tg\theta\frac{\sin\theta dr + r\cos\theta d\theta}{\cos\theta dr - r\sin\theta d\theta}}\right]d\theta = -r\frac{\sin^2\theta d\theta + \cos^2\theta d\theta}{\cos^2\theta dr + \sin^2\theta dr}d\theta \quad (67.4)$$

$$r = e^\theta \quad (67.5)$$

This is showing that the light in a point-mass universe is spiraling around the universe and against the Schwarzschild metric we don't arrive to any discontinuity and singularity.

In comparison to the Schwarzschild metric, we find a metric for light in a point-mass like universe with mass M that

$$dt^2 = \left(\frac{M}{\Diamond}\right)^2\left[\frac{dr^2}{r^2} + d\theta^2\right] \quad (68)$$

And this equation also is showing the spiral geodesy.

Of course for real universe from equations (33, 66) it is written

$$tg(\theta - arctg(dy/dx))d\theta = dC/C \quad (69)$$

In real condition, the universe has total Θ which is almost constant along the long cosmic times and then there is an almost constant k that

$$c = \frac{\Diamond}{k + m/r} \quad (70)$$

And then

$$tg(\theta - arctg(dy/dx))d\theta = \left(k + \frac{m}{r}\right)d\left(\frac{1}{k + \frac{m}{r}}\right) \quad (71.1)$$

$$\frac{tg\theta - \frac{\sin\theta dr + r\cos\theta d\theta}{\cos\theta dr - r\sin\theta d\theta}}{1 + tg\theta\frac{\sin\theta dr + r\cos\theta d\theta}{\cos\theta dr - r\sin\theta d\theta}}d\theta = \frac{m/r^2 dr}{(k + m/r)} \quad (71.2)$$

$$d\theta = \sqrt{\frac{m}{kr + m}}\frac{dr}{r} \quad (71.3)$$

This equation indicates a critical radius r_c that the light orbit comes into the spiraling motion when m/r is comparable to the universe M/R.

This critical radius is appeared while

$$\sum\frac{m}{r} \cong \frac{m}{r} \quad (72)$$

And then

BOOM!

$$r_c = \frac{G_N M}{c^2}$$

(73)

And this is about the same size which the Schwarzschild radius.

4 FUNDAMENTAL RAINBOW'S GRAVITY

The light refraction is a mystery yet. Newtonian gravity and Einstein general relativity and Maxwell electromagnetism can't answer to the question that why the light is refracted? It is not clear still why the light speed is slowed in the mediums. We don't know why the light is bent in the interface between the mediums.

We know that the explanations based upon the idea of light scattering from, or being absorbed and re-emitted by atoms are both incorrect. Explanations like these would cause a "blurring" effect in the resulting light, as it would no longer be travelling in just one direction. But this effect is not seen in the nature.

As mentioned in the Wikipedia dictionary, ultimate explanation for light refraction has been proposed on the nature of the light as an electromagnetic wave. On this explanation, the light is an oscillating electrical/magnetic wave. Then light traveling in a medium causes the electrically charged electrons of the material to also oscillate. Then it is thought that the oscillating electrons emit the electromagnetic waves of its own by this oscillation. It is thought that the emitted electromagnetic waves interact with the incident photons and when the waves interfere in this way, the resulting "combined' wave may have wave packets that pass an observer at a slower rate.

This explanation seems a joke first for interfere of the incident photons with produced electromagnetic waves by atoms. Mix of these waves is very imaginary and actually there is no such a chance. Even if it was a chance to interfere of these waves we were able to reproduce in the laboratory mix of the laser beams with an electromagnetic wave in the frequency, that of considered for emitted by atoms in the refraction whereas that we don't see it, that is, we don't observe a mix of the laser beams to slow the light speed as well as the refraction. Also if the wave packet was reason to decrease the speed of the light, then the white light which is also a wave packet had slower speed compared to the monochromatic beams. Second, the change of the angle is too a big problem in this explanation. There is a cheap explanation for that. The explanation states (Wikipedia) that:

"if it reaches the interface between the materials at an angle one side of the wave will reach the second material first, and therefore slow down earlier.

With one side of the wave going slower the whole wave will pivot towards that side. This is why a wave will bend away from the surface or toward the normal when going into a slower material. In the opposite case of as wave reaching a material where the speed is higher, one side of the wave will speed up and the wave will pivot away from that side."

The slowing of a part of the photon which reaches earlier to the different medium has no any mechanical source for change of the photon's direction in the interface of the mediums and disintegration of these fictitious wave packets to pure waves while the waves come out from the medium is also a paradox. Generally this explanation has no any argument for the Snell's law and slowing of the photon's speed in the mediums because of the group velocity of the wave packets has no any link to the Snell's law.

The mainstream physics has accepted that the light speed is not varied fundamentally in the refraction and mainstream scientists have accepted that the light speed variation is virtual, that is, what it is observed as the slowing of the light speed in the refraction is the group velocity.

In reality, the refractive index is an experimental index in the electromagnetism and there is no a mechanism to illustrate it and we don't know why the refractive index is varied in the different mediums electromagnetically and all are assumptions whereas that the Snell's law is very regular and such a law requires a verse origin.

The light is neutral electrically and then electromagnetic force can't refract the light. Also the Fermat least time and Huygens-Fresnel principles don't indicate the mechanical source for light speed variation in the refraction suppose these principles are technical lemmas and they don't show any mechanical potential for light refraction and its mechanism.

The argument for light refraction on the theory of the envelope mechanism and group velocity is mistake for many reasons for example the light is refracted when the light is even still not completely interred into the mediums. There should be a mechanical force affecting the photons from side of the atoms. The electromagnetic force can't be the case for the reality that the photons are mainly electrically neutral.

However the Newtonian gravity and Einstein general relativity can't answer to the light refraction but the gravitational potential as the integral of the masses per distances in fundamental gravity is not ever surrounded on the whole of the observable universe. As it is argued by Lutephy (2020), the gravitational potential in the nucleuses is limited to the wavelength of the force-carriers which naturally defines the range of the force which causes to increase highly the intensity of the gravity as the source of the strong nuclear force.

In fundamental gravity we have a mode which is weakest for that the potential is surrounded to the observable universe and a mode for the gravity which the integration of the potential is not on the whole of the observable universe but in the range of the wavelength of the quantum mechanical force-carriers. Then for a photons itself in the range of the photon's wavelength, photon plays the role of a force-carrier limited to the masses inside a sphere with radius equal to the photon's wavelength.

Thus, we have a system limited to a volume that the photon is itself quantum mechanically gravity-carrier between the particles in the range of the photon's wavelength and then a sphere with radius equal to the wavelength of the photon which photon is sited in the center is a quantum bound system as a quasi-universe for rainbow' gravity.

A failure for Newtonian gravity is that the G is constant whereas that according to the Mach inertia principle, the G is variable.

By the way from equation (50) for a photon's gravitational interaction in the range of the wavelength of the photon $r_i < \lambda$ we have

$$g = \frac{m_i c^2}{r_i^2 \sum\limits_{i | r_i < \lambda} \dfrac{m_i}{r_i}} \tag{74}$$

This equation is in reality generalization of the strong nuclear force which it has been argued by Lutephy (2020).

Equivalently we can write that

$$g = G_N \frac{\sum\limits_i \dfrac{m_i}{r_i}}{\sum\limits_{i | r_i < \lambda} \dfrac{m_i}{r_i}} \times \frac{m_i}{r_i^2} \tag{75}$$

This is the fundamental equation of the rainbow's gravity.

Then we have a generalized G in quantum bound system of the photons and generally the particles as

$$G = G_N \frac{\sum\limits_i \dfrac{m}{r}}{\sum\limits_{i | r_i < \lambda} \dfrac{m_i}{r_i}} \tag{76}$$

We can see that when the wavelength of the photon is equal to the radius of the observable universe, then the rainbow's gravity is at the same G, the newton's gravity so that $G = G_N$ whereas that how much the wavelength of the photon to be shorter, the G becomes larger so that for visible photons, the G becomes inasmuch as large that it curves the light intensely as well as the curvature of the light around the fictitious black holes.

Thus, the fundamental gravity in photonic quantum bound systems is reduced to the true rainbow's gravity. Of course according to the de Broglie wave equation $mv = h/\lambda$, each particle also has relevant quantum mechanical evolution via its wavelength which has been discussed for nucleuses previously (Lutephy, 2020).

By the way the complete equation of the rainbow's gravity is derived by fundamental gravity as

$$g = \frac{c^2}{\displaystyle\sum_{i|r_i<\lambda} \frac{m_i}{r_i}} \times \frac{m_i}{r_i^2}\left[\cos\varphi\vec{M} - \sin\varphi\vec{N}\right] \tag{77}$$

We see that when the wavelength is larger, the rainbow's gravity is weaker in a medium with constant number density. Then energetic photons are refracted much more.

On the equation (77) we can calculate the change of the light speed in N-body problem. This is not a regular process and it needs to resolve the apparatus of the equations. But from equation (77) approximately we find that Alpha Genesis Prime (Eq. 6) is generalized to the below equation included to a nonlinear term $\gamma_{r<\lambda}$ that:

$$c = \frac{\Diamond}{\displaystyle\sum_i \frac{m}{r} + \gamma_{r<\lambda}} \tag{78}$$

Where we have non-scale proportionality as

$$\gamma_{r<\lambda}\sum_{i\in r<\lambda}\frac{m_i}{r_i} = \gamma_0 \sum_{i\in r>\lambda}\frac{m_i}{r_i} \tag{79}$$

So that γ_0 is assumed as a constant to justify the equation.

Then in the rainbow's gravity, the linear equation of the Alpha Genesis Prime is not valid longer.

Interestingly we find an approximation so that

$$\gamma_{r<\lambda} \approx \sum_i \frac{m_i}{r_i^{\alpha(\lambda)}} \tag{80}$$

This equation fulfills both the properties, diminishing the term for $r > \lambda$ and enhanced G for smaller wavelengths and then we have

$$c = \frac{\Diamond}{\sum_i \frac{m}{r} + \sum_i \frac{m_i}{r_i^{\alpha(\lambda)}}} \tag{81}$$

On this nonlinear variable light speed in the rainbow's gravity mode which is derived by fundamental gravity, we can write a metric so that

$$ds^2 = -c_0^2 \left(1 + \frac{G_N m}{c_0^2 r^{\alpha(\lambda)}}\right)^{-1} dt^2 + \left(1 + \frac{G_N m}{c_0^2 r^{\alpha(\lambda)}}\right) dr^2 + r^2 d\Omega^2 \tag{82}$$

In a different method the scientists (Hendi et al., 2016) have extracted a rainbow's metric for spherical symmetric space-time in d-dimensions as

$$ds^2 = -\frac{\psi(r)}{f(E)^2} dt^2 + \frac{1}{g(E)^2} \left[\frac{dr^2}{\psi(r)} + r^2 d\Omega^2\right] \tag{83}$$

Where according to the (Hendi et al., 2016):

$$d\Omega^2 = d\theta_1 + \sum_{i=2}^{d-2} \prod_{j=1}^{i-1} \sin^2 \theta_j d\theta_i^2 \tag{84}$$

A solution is Reissner-Nordstrom black hole solution (Hendi et al., 2016) as

$$\psi(r) = 1 - \frac{m}{r^{d-3}} - \frac{2Ar^2}{(d-1)(d-2)} + \frac{2(d-3)q^2}{(d-2)r^{2(d-3)}} \tag{85}$$

The scientists have used d-dimensions to find such solutions to agree with the cosmic observations of the gamma bursts. But by comparison of the equations (82, 83), if we assume that

$$\psi(r) = \left(1 + \frac{G_N}{c_0^2} \frac{m}{r^{\alpha(\lambda)}}\right) \tag{86}$$

Then we deduce that

$$f(E) = g(E) = 1 \tag{87}$$

Then in comparison with the equation of modified dispersion relation (MDR) we obtain that

$$E^2 - p^2 = m^2 \tag{88}$$

And we understand that the dependency of the light speed to the photon wavelength doesn't change the energy-momentum equation (Eq. 88).

We find here that the generalization of the general relativity to the quantum bound systems can be treated on the $\psi(r)$ only. For example when we assume that $\alpha(\lambda) = 3$, then for universe matter we find that $\Theta_U \equiv 10^{-25}$ and then the effect of the universe potential almost diminishes and it is appeared a high amplitude micro-gravity in general relativity.

Or if we assume that $\alpha(\lambda) = 4$, then the metric tensor (Eq. 82) shows the

strong nuclear force limited to the nucleuses in general relativity.
Now we find perfect formula of refractive index, that is,

$$cn = c_0 \qquad (89.1)$$

$$n = \frac{\sum_i \dfrac{m_i}{r_i} + \gamma_{r<\lambda}}{\sum_i \dfrac{m_i}{r_i}} \qquad (89.2)$$

$$n \cong 1 + \frac{G_N}{c_0^2} \sum_i \frac{m_i}{r_i^{\alpha(\lambda)}} \qquad (89.3)$$

We find that the rainbow's gravity is related to a nonlinear potential term which causes to appear dispersion relations in the gravity.

By the way we find here that the refraction is happened even out of the material bar in the range of the wavelength of the incident photons. This means that if a laser beam to pass around a material bar, the refraction will be done. Such an experiment was performed previously by Mahmoud Hessaby and he realized that the laser beams are refracted by material bar and then Mahmoud Hessaby (1947, 1948) for correlation of the refractive index to the volume density theorized the theory of the extended particles.

Of course the Hessaby's report for laser beam bending around a material bar mistakenly has been considered as the Gaussian diffraction whereas that the diffraction is symmetric and the Hessaby's report was indeed the light refraction by material bar as we see that it is related to the rainbow's gravity. Hessaby's mistake in the interpretation of the phenomenon was that he took into account the correlation of refractive index to the volume density whereas that such a correlation is not fundamental. In fact the volume density is not fundamental for reality that the absolute density is not mass embedded in a volume but it is formulized by integral of the masses per distances from the point we want to calculate the absolute density there. The true density in the fundamental gravity is absolute density (potential) and what the scientists have considered as the mass embedded in the volume is an approximation. There is no any volume fulfilled uniformly by matter and simply we find falsification for volume density in the fundamental physics. Each object is included to the discrete atoms in a region of the space and definition of the volume density for light traveling between the atoms is meaningless.

Against the theory of particles infinity by Mahmoud Hessaby (1947, 1948), the density is overall in definition of the absolute density and we see that the refraction is not closed fundamentally to the volume density but refraction is related to the absolute density, of course possible to transfer to the approximate equations in the definition of volume density.

Of course however Hessaby's theory seems incorrect but his report for light bending out of the material bar is correct and his report for light refraction by material bar is really true and maybe fit for noble prize in that time. Also we should notice that we don't deny the extended particle theory but we find that the refraction is not by extended particles even if the particles are extended. In fact the quantum mechanics is a kind of extended particle theory. Difference is that in extended-particle theory, the mass is integral of the energy of extended-particle on the whole of the space whereas that in the quantum mechanics, the integral of the existence probabilities is at the order of the unity on the whole of the space.

In the Einstein general relativity, the curvature is proportional to the energy tensor and then general relativity requires extended-particle energy which fulfills the continuity of the equations. In Einstein general relativity, the space should be fulfilled with the energy and this leads to the extended definition of the energy. Then the continuum and extended-particle theory of the Mahmoud Hessaby is in reality complementary of the Einstein general relativity.

5 GRAVITY AS THE POTENTIAL SOURCE OF THE LIGHT DIFFRACTION

As we know from literature, the diffraction in definition is the slight bending of the light as it passes the edge of an object. And amount of the bending depends on the relative size of the wavelength of light to the size of opening. If the opening is much larger than the light's wavelength, the bending will be almost unnoticeable. However, if the two are closer in size or equal, the amount of bending is considerable. We know that the characteristic bending pattern is almost pronounced when a wave from coherent source encounters a slit/aperture that is comparable in size to its wavelength.

The science still has no any answer to the question that

What is the source of light bending in diffraction?

The scientists have proposed a mechanism that the optical effects resulting from diffraction are produced through the interference of light waves which are propagated from the secondary wave sources on the Huygens-Fresnel principle. But this is not the answer for the question that what is the source of potential for light bending and the Huygens-Fresnel principle doesn't answer the question and doesn't offer any alternative mechanical potential and in reality Huygens-Fresnel principle is a shortcut model. Huygens-Fresnel principle however yields to the equations for the Fraunhofer and Fresnel fields, but it doesn't reveal the source of the potential of light bending in the diffraction and the enigma is unsolved still. But here on the fundamental rainbow's gravity we see that while the wavelength of the photon is comparable to the size of the slit/aperture, the slit/aperture becomes a bound quantum system (BQS). Of course while the wavelength of the photon is longer than the scale of the opening, the aperture bounces off it. Then slit/aperture is a gravitational bound quantum system when the wavelength of the incident photons is comparable to its opening scale. This is reasonable to bend the photons by fundamental rainbow's gravity. The rainbow's gravity does bend the photons while the photons pass through the slit/aperture like the light refraction. In fact the refraction and diffraction are identical in potential source.

In the diffraction, obstacle is reasonable to create secondary wave sources by the atoms which the incident photons are passing through it whereas that in the light refraction, the photons pass among the atoms.

While a slit/aperture length scale is comparable to the wavelength of the incident photons, the slit/aperture causes to distribute the photons coherent in the mean of a secondary wave source. Of course in one silt experiments, the edges of the opening play the role of secondary wave sources and it is possible to obtain a qualitative understanding of many diffraction phenomena by considering how the relative phases of the individual secondary wave sources vary.

We need to notice that the Huygens-Fresnel principle is a shortcut model to illustrate the diffraction but exact answer of the diffraction should be based on the N-body problem of the distributed photons under rainbow's gravitational potential between the photons in the range of the wavelength of the photons and this problem is a quantum mechanical approach.

Huygens-Fresnel principle yields to the superposition of the waves whereas that in the quantum approaches, the diffraction pattern is created by the distribution of paths, the observation of light and dark bands is the presence or absence of photons in these areas. As we know there are various analytical models which allow the diffracted field to be calculated, including the Kirchhoff-Fresnel diffraction equation which is derived from the wave equation (Baker and Copson, 1939).

By the way the diffracted photons don't move straight suppose however the Huygens-Fresnel matches the reality but the path of the diffracted photons is curved. This reality verifies that the diffracted photons interact each other mechanically and then it needs a potential source between the diffracted photons, that is, the rainbow's gravity. In the absence of the rainbow's gravity, the path of the photons was ever a straight path. This means that the diffraction is not driven just by the obstacles but the obstacles cause to distribute the photons to interact each other by the rainbow's gravity in full time of the path. Then the diffraction is not resolved like the Schrodinger wave equation for a particle in a box but it needs to resolve it for a self-focusing potential between the photons distributed by the sources positioned in the edges of the slit/aperture. Clearly this problem is almost impossible to resolve mathematically but fortunately the Huygens-Fresnel principle is a good approximation as a mathematical model to find the dark and bright pattern of the diffraction.

6 GRAVITY AS THE POTENTIAL SOURCE OF THE PHOTOELECTRIC AND GENERALLY THE PHOTONS AND MATTER INTERACTION

On the Einstein research (1916), the interaction between the mater and electromagnetic radiation can be illustrated on the three processes, the inductive radiation and stochastically radiation and inductive absorption. Up to now, interaction between the matter and photons has been discussed but the research papers don't focus to the potential source of the interaction between the matter and photons. Yet we don't know why the photons are interacted by the atoms and electrons and why the inductive and stochastically radiations and absorptions are happened. We don't know the potential source of the thermodynamics in the level of the interaction between the particles and photons.

The mainstream mechanism is based on the proof less principles and wave solutions, but not focused to the potential source of the events.

For example according to the Einstein illustration, the inductive absorption is happened while the photons with energy $h\nu = E_1 - E_0$ reach the matter with base energy E_0 and excitation energy E_1. But neither the Einstein nor the next scientists have mot focused to the potential source of the photons and matter interaction. They think that the interaction is electromagnetic whereas that the photons are electricity neutral.

For photons scattering the scientists have found some equations like the Rayleigh scattering law (Feynman et al., 1963) and even the more general theory of the scattering has been developed, so called Mie scattering (Moyer-Arendt, 1989) but still the potential source of the interaction between the photons and particles is unknown.

But here we answer it simply. We use also from simplest model.

If there is a system included to a photon and a particle, if the distance between them is smaller than the wavelength of the photon then from equation (77) we have

$$\left| g \right| = \frac{c^2}{r}$$

(90)

Then according to the Newton's second law we obtain

$$\frac{c^2}{r} = \frac{cd\theta}{dt} = \frac{c^2}{R}$$

(91.1)

$$r = R$$

(91.2)

This means that the photon is absorbed by the particle via rainbow's gravity as the mechanical source of the photon absorption by particles. Scientists are thinking that such a gravitational absorption is possible via the mini black holes which it needs energy of the order of the Planck energy ($\sim 10^{19} GeV$) which is beyond what can be achieved in the near future. But here we see that the rainbow's gravity in the range of the micro gravity is inasmuch as big that it simulates the black holes by atoms and photons, what the scientists are looking for it in the colliders in large extra dimensions as noted by Ali et al., (2015) that:

"This is because the existence of large extra dimensions can lower the effective Planck scale to TeV scales at which experiments can be done (Arkani-Hamed et al., 1998)."

Then the rainbow's gravity is the potential source of the interaction between the photons and atoms.

In reality the scattering and diffraction are identical in the source. The absorption and radiation by matter is a statistical process driven by the population of the photons interactive with atoms via the rainbow's gravity and the photoelectric is an envelope moving with group velocity driven by the electron and interactive photon. Then on the thermodynamic equilibrium, the radiation and absorption are at the same order and random radiation of the matter is driven by the fact that the interaction between the photons and matter is statistical and uncertain.

Here the illustration of the interaction between the photons and matter is not mainly different but we find a potential source for what the scientists have explained by some principles and lemmas. Of course the photons are affected slightly by electric field too for that the photons are additionally neutral but included to the electrically positive and negative strings.

7 GRAVITY ON THE MOVING BODIES AND NEWTONIAN GRAVITY AS AN APPROXIMATION

While the fundamental gravity is written in the Newtonian force mode which is a vector divided to the tangent and normal components on the light velocity. We find that the gravity is not a centrifugal force but it is approximately centrifugal for static mass, verifying that in Newtonian mode the static mass is not fundamental but associated motion of the fundamental particles is virtually static. In fact philosophically there is no any fundamental station in the universe and according to the Alpha Genesis Prime; the fundamental particles are in their maximum velocities which is determined by the Alpha Genesis Prime.

Everything is dynamic and the station is composite. This is confirmed also with the equivalency of the mass and energy so that when we can transfer the static mass to the photon, then the static mass is essentially dynamic. The fundamental gravity does force on the light speed principally but the moving bodies are not principal suppose composite of the principal fundamental particles in their fundamental velocities.

By the way, if the speed of mass is being zero we can state that

<u>In the same mass, sum of the motions is zero.</u>

In the ideal formation a static mass is a ring but a light in ideal formation is a string.

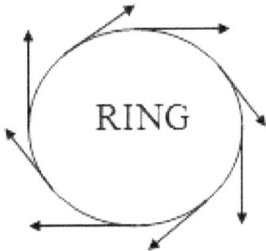

Figure 3. simple ring to show the station of additional motion included to the symmetric fundamental velocities.

A ring in addition has no any direction why that the ring is rotating and then the speed of ring is zero and the symmetry in the ring causes that

$$\sum_i \vec{c}_i = 0 \tag{92}$$

Then the integration of tangential components of the forces on the ring is zero, that is,

$$\sum_i f_{Mi} = 0 \tag{93}$$

Then the gravity on the ring is equal to the integration of normal components.

For symmetry of the normal components about the vector $\vec{r}$ it is resulted

$$f_{ring} = \sum_i f_{Ni} = -\frac{Gm_i}{r_i^2} \tag{94}$$

And then newton's gravity is obtained as an approximation from fundamental gravity for static mass.

This argument shows that for Newton's gravity it isn't necessity that the moving bodies with velocities lesser than the maximum velocity to be fundamental. We can state that the fundamental velocities are maximum velocities for that the composition decreases the velocity. In fact the fundamental velocity is an upper limit for composite velocities. There are many evidences for this paradigm but here we limit the subject on the gravity only and not discussion in detail the structure of the particles.

8 LOGICAL CONTRADICTIONS IN GTR FOR BLACK HOLE AND LIGHT BENDING BY SPACE CURVATURE

8.1 Larger bending of the light by the Sun isn't supported by the curvature of Space

The Schwarzschild metric (Schwarzschild, 1916) for a planar geodesic is

$$ds^2 = c_0^2\left(1-\tfrac{\alpha}{r}\right)dt^2 - \frac{1}{\left(1-\tfrac{\alpha}{r}\right)}dr^2 - r^2 d\theta^2 \tag{95}$$

When we assume $\omega = 1 - \tfrac{\alpha}{r}$, then resolution of this metric results an apparatus of three equations that

$$0 = \frac{d^2\theta}{d\tau^2} + \frac{2}{r}\frac{d\theta}{d\tau}\frac{dr}{d\tau} \tag{96.1}$$

$$0 = \frac{d^2 t}{d\tau^2} + \frac{1}{\omega}\frac{d\omega}{dr}\frac{dt}{d\tau}\frac{dr}{d\tau} \tag{96.2}$$

$$0 = \frac{d^2 r}{d\tau^2} - \frac{1}{2\omega}\frac{d\omega}{dr}\left(\frac{dr}{d\tau}\right)^2 - r\omega\left(\frac{d\theta}{d\tau}\right)^2 + \frac{c^2\omega}{2}\frac{d\omega}{dr}\left(\frac{dt}{d\tau}\right)^2 \tag{96.3}$$

We know from solution of the Eq. (96.2) that

$$\frac{dt}{d\tau} = \frac{k}{\omega} \Big| k = const \tag{97}$$

And in the literature it is used below approximations to resolve the equation (97.3):

$$\omega = 1 \tag{98.1}$$

$$\frac{d^2 t}{d\tau^2}\frac{dr}{dt} = 0 \tag{98.2}$$

By these approximations we obtain from Eq. (96.3) that

$$\frac{d^2 r}{dt^2} = -\frac{\alpha c^2}{2r^2}\left(1-(v/c)^2\right) \tag{99}$$

This relation violates Einstein equivalence principle. However, the tensor components of Schwarzschild metric are velocity-independent but equation of the motion on the Schwarzschild metric yields to velocity-dependent equation. Of course to surpass this paradox, in the literature it is used the proper time τ instead the time t and on the base of Eq. (97) it is deduced

$$\frac{d^2 r}{dt^2} = \left(\frac{\omega}{\gamma}\right)^2 \frac{d^2 r}{d\tau^2}\bigg| k = \gamma = \frac{1}{\sqrt{1-\left(\frac{v}{c}\right)^2}} \qquad (100)$$

And by assumption of the Eq. (100) and embedding it into the Eq. (99) it is deduced that

$$\frac{d^2 r}{dt^2} = -\frac{\alpha c^2}{2r^2} \qquad (101)$$

And by so-called relativistic definition $\alpha = 2GM/c^2$, the Eq. (101) is the same Newtonian gravity. In reality the condition $\omega \sim 1$ yields to the gravity at the infinity, that is, the assumption that the gravity in the infinity is Newtonian in the general relativity. Then it seems that the paradox is resolved from Einstein general relativity but the Eq. (101) is clearly wrong because that it isn't compatible with relativistic light speed $c = c_0 \omega$ in Schwarzschild metric and differentiation on the light speed in Schwarzschild metric is showing that

$$\frac{d^2 r}{dt^2} = \frac{dc}{dt} = c\frac{dc}{dr} = +\frac{2GM}{r^2} \qquad (102)$$

And this equation does show a clear contradiction.

Then by transformation of time t to the proper time τ, again deduced answer is false whereas that the Eq. (99) is invariant under proper time and proper length transformations that

$$\begin{cases} a' = a/\gamma^2 \\ r' = r\gamma \end{cases} \qquad (103)$$

Where the proper acceleration is symbolized by a' and proper distance with r'.

On the other hand, the tensors are invariant under frame transformation and the answer in the coordinates (r, t) is adequate. Even, the relation $dt = \gamma d\tau$ is wrong because that the proper time is not the time at the frame we use Lorentz transformation suppose the proper time is invariant in all the frames and then the assumption $k = \gamma$ in the Eq. (97) is a null hypothesis.

In exact resolution if we want to use from the proper time instead the time t, we need to use from below complete equation instead the approximate Eq. (100) so that

$$\frac{d^2r}{d\tau^2} = \frac{d^2r}{dt^2}\left(\frac{dt}{d\tau}\right)^2 + \frac{d^2t}{d\tau^2}\frac{dr}{dt} \qquad (104)$$

And applying Eq. (104) instead approximate Eq. (100) and repeating the calculations results

$$\frac{d^2r}{d\tau^2} = -\frac{c^2}{2\omega^2}\frac{d\omega}{dr}\left(1 + 2\gamma(v/c)^2\right) \qquad (105)$$

Then applying proper time instead the time t, again it violates the equivalence principle.

On the other hand, the Eq. (99) is not the answer of the Schwarzschild metric for that the assumptions in the Eq. (98.1, 98.2) are approximate whereas that on the Schwarzschild metric we have

$$\begin{cases} \omega = 1 - \dfrac{\alpha}{r} \\ \dfrac{d^2t}{ds^2}\dfrac{dr}{dt} \neq 0 \end{cases} \qquad (106)$$

Then we need to use below complete equation that

$$\frac{d^2r}{d\tau^2} = \frac{d}{d\tau}\left(\frac{dr}{d\tau}\right) = \frac{d}{d\tau}\left(\frac{dr}{dt} \times \frac{dt}{d\tau}\right) = \frac{d^2r}{d\tau dt}\frac{dt}{d\tau} + \frac{dr}{dt}\frac{d^2t}{d\tau^2} = \frac{d^2r}{dt^2}\left(\frac{dt}{d\tau}\right)^2 + \frac{d^2t}{d\tau^2}\frac{dr}{dt} \qquad (107)$$

By this equation, the equation (96.3) is written as

$$\frac{d^2r}{dt^2} + \frac{d^2t}{d\tau^2}\frac{dr}{dt}\left(\frac{d\tau}{dt}\right)^2 = \frac{1}{2\omega}\frac{d\omega}{dr}\left(\frac{dr}{dt}\right)^2 - \frac{c^2\omega}{2}\frac{d\omega}{dr} + r\omega\left(\frac{d\theta}{dt}\right)^2 \qquad (108)$$

Embedding Eq. (97.2) in this equation implies

$$\frac{d^2r}{dt^2} = \left(1 + \frac{1}{2\omega}\right)\frac{d\omega}{dr}\left(\frac{dr}{dt}\right)^2 - \frac{c^2\omega}{2}\frac{d\omega}{dr} + r\omega\left(\frac{d\theta}{dt}\right)^2 \qquad (109)$$

$$\frac{d^2r}{dt^2} = -\frac{c^2}{2}\frac{d\omega}{dr}\left[\omega - \left(2 + \frac{1}{\omega}\right)(v/c)^2\right] + r\omega\left(\frac{d\theta}{dt}\right)^2 \qquad (110)$$

In a polar coordinate system, it is deduced by vector calculus that the acceleration in space is

$$\vec{a} = \left(\ddot{r} - r\dot{\theta}^2\right)\hat{r} + \left(2\dot{r}\dot{\theta} + r\ddot{\theta}\right)\hat{\theta} \qquad (111)$$

The $\ddot{r}$ is radial acceleration and $r\dot{\theta}^2$ is centrifugal acceleration and $r\ddot{\theta}$ is angular acceleration and $2\dot{r}\dot{\theta}$ is Coriolis acceleration. Interestingly on the Schwarzschild metric it is deduced that

$$\left(2\dot{r}\dot{\theta}+r\ddot{\theta}\right)=0 \tag{112}$$

And then from equations (112) and (111) it is deduced

$$\vec{a}=\left(\ddot{r}-r\dot{\theta}^2\right)\hat{r} \tag{113}$$

By comparison of the equations (110) and (113) it is deduced

$$\vec{a}=\frac{d^2\vec{r}}{dt^2}=-\frac{c^2}{2}\frac{d\omega}{dr}\left(\omega-\left(2+\frac{1}{\omega}\right)(v/c)^2\right) \tag{114}$$

Then we have

$$\frac{d^2\vec{r}}{dt^2}=-\frac{c^2}{2}\frac{d\omega}{dr}\left(\omega-\frac{3}{\omega}(v/c)^2\right) \tag{115}$$

By this relation we obtain

$$\frac{d^2\vec{r}}{dt^2}=-\frac{c^2}{2}\frac{\alpha}{r^2}\left(\omega-\frac{3}{\omega}(v/c)^2\right) \tag{116}$$

Then if we assume like Schwarzschild (1916) that $\alpha=2GM/c^2$ we obtain

$$\frac{d^2\vec{r}}{dt^2}=-\frac{GM}{r^2}\left(\omega-\frac{3}{\omega}(v/c)^2\right) \tag{117}$$

The Eq. (117) is quasi-Newtonian and even invariant by the proper Lorentz transformation. Then the inertial mass per gravitational mass in Schwarzschild metric is deduced as

$$\frac{m_i}{m_g}=\left(\omega-\frac{3}{\omega}(v/c)^2\right) \tag{118}$$

And this relation does violate the equivalence principle and also by Eq. (117) it is deduced that

$$\begin{cases} v=c \Rightarrow \dfrac{d^2r}{dt^2}=-\left(\omega-\dfrac{3}{\omega}\right)\dfrac{GM}{r^2}\approx+\dfrac{2GM}{r^2} \\[2mm] v=\omega\dfrac{c}{\sqrt{3}} \Rightarrow \dfrac{d^2r}{dt^2}=0 \\[2mm] v=0 \Rightarrow \dfrac{d^2r}{dt^2}=-\dfrac{GM}{r^2}\omega \end{cases} \tag{119}$$

Then relativistic larger light bending around the Sun sourced by space curvature is a big mistake but the equations are showing that the light inertial mass is ever ½ compared to the static mass and this effect is not dependent to the space curvature, suppose independent of the distance r. We see that the Einstein field results a critical velocity that

$$v_e = \omega c / \sqrt{3} \tag{120}$$

Then it is deduced that

$$\begin{cases} v < v_e \Rightarrow a < 0 \\ v = v_e \Rightarrow a = 0 \\ v > v_e \Rightarrow a > 0 \end{cases} \tag{121}$$

Then out of the Schwarzschild radius ($\omega > 0$), Einstein field vacuum solution implies repulsive force for speeds larger than critical velocity and the acceleration is zero on the critical velocity of the moving bodies. Scientists have used direct calculations on the Schwarzschild metric and generating (E=total energy, L=angular momentum) that

$$\left(\frac{dr}{d\theta} \right)^2 = \frac{r^4}{b^2} - \omega \left(\frac{r^4}{a^2} + r^2 \right) \Bigg| \begin{cases} a = \frac{L}{mc} \\ b = \frac{Lc}{E} \end{cases} \tag{122}$$

The light bending by Eq. (122) is equal to $\delta\phi_E \sim 4GM / c^2 b$ and when we apply the newton mechanics, we obtain the light bending as $\delta\phi_N \sim 2GM / c^2 b$.

Then also by Eq. (122) we find that the relativistic larger bending of the light is independent of the Sun's mass and the Sun's distance, but the light inertia is ½ compared to the static mass. Even at the infinity, where the curvature of the space is zero, we see that $\delta\phi_E / \delta\phi_N = 2$. Then the space curvature around the sun (where $\omega \sim 1$) is not the source of light larger bending and for every active mass $(\omega \sim 1)$ instead the Sun, the light bending would be almost twice as great as the Newtonian gravity. Topologically every metric is inconsistent with complete attractive acceleration suppose for each metric when the light speed is increased by distance, there is a critical speed that the acceleration for moving bodies with rather speeds is repulsive. To argue consider an arbitrary metric which the light speed is increasing radially like the Schwarzschild metric. It is possible a moving body with speed v which along the time reduces to speed zero and no come back. Then there is a critical speed that the acceleration is zero.

8.2 No black hole supported by the Einstein general relativity

By Eq. (117), for bodies moving with velocity larger than the critical velocity, out of the Schwarzschild metric, the force is attractive but for bodies moving with velocities larger than the critical velocity, out of the Schwarzschild radius, the force is repulsive and then against the definition of the black hole, the bodies with velocities larger than the critical velocity never to inter inside of the Schwarzschild radius. To inter it requires below condition to be agreement:

$$\omega - \frac{3}{\omega}\left(\frac{v}{c}\right)^2 > 0 \tag{123}$$

Then for light to inter in a black hole it should be ever agreement that

$$\omega > \sqrt{3} \tag{124}$$

But this is impossible and light escapes form the surface of the Schwarzschild radius in an increasing velocity. In reality, where at the Schwarzschild radius, the light velocity is zero, then the light cant inter inside the Schwarzschild radius unless via quantum and if entrance to be quantum then escape is quantum too and this phenomenon yields to a thermodynamic equilibrium and then never a black hole established.

But for inside the Schwarzschild radius, we see that the light speed is negative and we don't understand how we can explain physically, the mathematical negative velocity of the light in general relativity. By the way, if we don't notice to this paradox, we see that the light speed is increasing in an absolute size when the radius is decreased and light speed becomes infinity at the center ideally. By critical velocity $v_e = \omega c / \sqrt{3}$, inside the Schwarzschild radius, we have

$$\begin{cases} v < v_e \Rightarrow a > 0 \\ v = v_e \Rightarrow a = 0 \\ v > v_e \Rightarrow a < 0 \end{cases} \tag{125}$$

This means that the gravity inside the Schwarzschild radius is repulsive for the bodies moving with velocity lower than the critical velocity and then amazingly what we have called it the black hole, it repulses the matter where $|v| < |\omega C| / \sqrt{3}$ and this is a complete violation for relativistic black hole. Inversely inside of the Schwarzschild radius we find vacuum hole.

Even where $|v| > |\omega C| / \sqrt{3}$ inside the Schwarzschild radius, the bodies would move to the side of the Schwarzschild radius to remain there most of the time for that the velocity tends to zero.

Then we understand that the matter with velocities lower than the critical velocity would be pressed at the Schwarzschild radius generating high temperature shell for that out of the Schwarzschild radius, the force is attractive for bodies moving with velocity lower than the critical velocity and for inside the Schwarzschild radius, the force is repulsive for bodies moving with velocity lower than the critical velocity. Then what we find from Schwarzschild metric is a spherical high temperature bright shell, not a black hole. Interestingly, our results are verified by results derived previously. As found in papers (Carmeli, 1972; Blinnikov et al., 2000, 2001), in the gravitational field of a spherically symmetric object, there exists a critical value of the coordinate speed $v_c = c/\sqrt{3}$.

References

Ali A.F., (2015). Absence of black holes at LHC due to gravity's rainbow. Physics Letters B 743 295–300.

Ali A.F., M.M. Khalil, (2014). A proposal for testing gravity's rainbow, eprint number: 1408.5843.

Ali A.F., M. Faizal., B. Majumder, (2015). Absence of an effective horizon for black holes in gravity's rainbow, Europhys. Lett. 109, 20001, http://dx.doi.org/10.1209/0295-5075/109/20001.

Amelino-Camelia, Giovanni, et al., (2011). principle of relative locality. Physical review D 84.8: 084010

Amelino-Camelia, G., J.R. Ellis, N. Mavromatos, D.V. Nanopoulos, (1997). Distance mea-surement and wave dispersion in a Liouville string approach to quantum grav-ity, Int. J. Mod. Phys. A 12, 607–624.

Amelino-Camelia, G., J.R. Ellis, N. Mavromatos, D.V. Nanopoulos, S. Sarkar, (1998). Tests of quantum gravity from observations of gamma-ray bursts, Nature 393, 763–765.

Antoniadis I.,, N. Arkani-Hamed, S. Dimopoulos, G. Dvali, (1998). New dimensions at a millimeter to a Fermi and superstrings at a TeV, Phys. Lett. B 436, 257–263.

Arkani-Hamed N, S. Dimopoulos, G. Dvali, (1998). The Hierarchy problem and new dimensions at a millimeter, Phys. Lett. B 429, 263–272.

Awad A., A.F. Ali, (2013). Majumder B. Nonsingular rainbow universes, J. Cosmol. As-tropart. Phys.1310, 052.

Banks T., W. Fischler, (1999). A model for high-energy scattering in quantum gravity, Report number: RU-99-23, UTTG-03-99, eprint number: hep-th/9906038.

Baker B.B., and E.T. Copson, (1939). The mathematical theory of Huygens's principle. Clarendon Press, Oxford, U.K.

Barrow J.D., J. Magueijo, (2013). Intermediate inflation from rainbow gravity, Phys. Rev. D 88, 103525.

Blinnikov, S. I., L.B. Okun, M.I. Vysotsky, (2000, 2001). arXiv: gr-qc/0111103; Proc. of the Workshop "What comes beyond the Standard Model 2000, 2001", H. Nielsen Festschrift, ed. by N.M. Borstnik, C.D. Froggatt, D. Lukman, vol.2, p. 116.

Carmeli M, 1972: Lett. Nuovo Cimento **3**, 379.

Einstein A, (1916). The foundation of the general theory of relativity. Annalen der physik **49** (7): 769-822

Chatrchyan, S et al., (2012). CMS Collaboration, Search for dark matter and large extra dimensions in monojet events in ppcollisions at $\sqrt{s}$=7TeV, J. High Energy Phys. 1209 (2012) 094.

Chatrchyan, S et al., (2012). CMS Collaboration, Search for microscopic black holes in ppcollisions at $\sqrt{s}$=7TeV, J. High Energy Phys. 1204, 061.

da Rocha R., C.H. Coimbra-Araujo, (2006). Extra dimensions in LHC via mini-black holes: effective Kerr–Newman brane-world effects, Phys. Rev. D 74, 055006.

Dimopoulos S., G.L. Landsberg, (2001). Black holes at the LHC, Phys. Rev. Lett. 87, 161602.

Emparan R., G.T. Horowitz, R.C. Myers. (2000). Black holes radiate mainly on the brane, Phys. Rev. Lett. 85, 499–502.

Feynman, Richard P., Robert B. Leighton, and matthew Sands. (1963). The Feynman Lectures of physics, Vol. 1. Reading, Mass.: Addison-Wesley Publishing Compan. Ch. 32, 33.

Feng, Zhong-Wen and Yang, Shu-Zheng. (2018). Rainbow gravity corrections to the entropic force. Advances in high energy physics. Vol. 2018.ticle ID 5968284, 8pages. https://doi.org/10.1155/2018/5968284

Galan P., G.A. Mena Marugan. (2004). Quantum time uncertainty in a gravity's rainbow formalism, Phys. Rev. D 70, 124003.

Garattini R., G. Mandanici. (2012). Particle propagation and effective space–time in gravity's rainbow, Phys. Rev. D 85, 023507.

Garattini R., B. Majumder. (2014). Naked singularities are not singular in distorted grav-ity, Nucl. Phys. B 884, 125–141.

Garattini R., (2013). Distorting general relativity: gravity's rainbow and $f(R)$ theories at work, J. Cosmol. Astropart. Phys.1306, 017.

Garattini R., B. Majumder. (2014). Electric charges and magnetic monopoles in gravity's rainbow, Nucl. Phys. B 883, 598–614.

Giddings S.B., S.D. Thomas, (2002). High-energy colliders as black hole factories: the end of short distance physics, Phys. Rev. D 65, 056010.

Gim Y., W. Kim, (2014). Thermodynamic phase transition in the rainbow Schwarzschild black hole, J. Cosmol. Astropart. Phys.10, 003.

Hackett J, (2006). Asymptotic flatness in rainbow gravity, Class. Quantum Gravity 23, 3833–3842.

Hendi, S.H., B. Eslam Panah., S. Panahiyan., M. Momennia, (2016). F(R) gravity's rainbow and its Einstein counterpart. Advances in High Energy Physics. Vol. 2016, Article ID 9813582. http://doi.org/10.1155/2016/9813582

Hessaby M, (1947). "continuous particles". Proceedings of the national academy of scineces of the united states of America. 33 (6): 189-194.

Hessaby M, (1948). "theoretical evidence for the existence of a light-charged particle of mass greater than that of the electron". Phys. Rev. 73 (9): 1128

Leiva C., J. Saavedra, J. Villanueva, (2009). The geodesic structure of the Schwarzschild black holes in gravity's rainbow, Mod. Phys. Lett. A 24 1443–1451.

Li H., Ling Y., Han X, (2009). Modified (A)dS Schwarzschild black holes in rainbow spacetime, Class. Quantum Gravity 26. 065004.

Liu C.-Z., Zhu J.-Y, (2008). Hawking radiation and black hole entropy in a gravity'srain-bow, Gen. Relativ. Gravit. 40, 1899–1911.

Lutephy M, (2019). Explosion of the light: Alliance of the Planck and Mach and Einstein-Riemannian physics. Physics International, 10(1), 8-23. http://doi.org/10.3844/pisp.2019.8.23

Lutephy M, (2020). Explosion of the Science: Modification of the Newtonian dynamics via Mach's inertia principle and generalization in gravitational quantum bound systems and finite range of the gravity-carriers, consistent merely on the bosons and the fermions. Physics International, 11(1), 4-35. http://doi.org/10.3844/pisp.2020.4.35

Lutephy M, (2019). Absolute physics: non scale mechanics. ISIN: 1492164216.

Meade P., L. Randall, (2008). Black holes and quantum gravity at the LHC, J. High En-ergy Phys. 0805, 003.

Moyer-Arendt, Jurgen R, (1989). Introduction to classical and modern optics, 3d ed. Engle-wood Cliffs, N.J.: Prentice-Hall, Ch. 4.2.

Schwarzschild K, (1916). „"Uber das gravitationsfeld eines massenpunktes nach der einsteinschen theorie"". Sitzungsberichte der koniglich preussischen akademie der wissenschaften 7: 189-196.

BOOM!

ABOUT THE AUTHOR

I am a physician and mathematician and chemistry engineer and generally a scientist and I have published the papers and the books and I have many discoveries consequently I will publish them. Please email me at: lutephy@gmail.com

BOOM!

www.ingramcontent.com/pod-product-compliance
Lightning Source LLC
Chambersburg PA
CBHW070515220526
45467CB00002B/680